RACCOON

COLORING BOOK FOR ADULTS

Copyright: Published in April
the United States

All rights reserved.

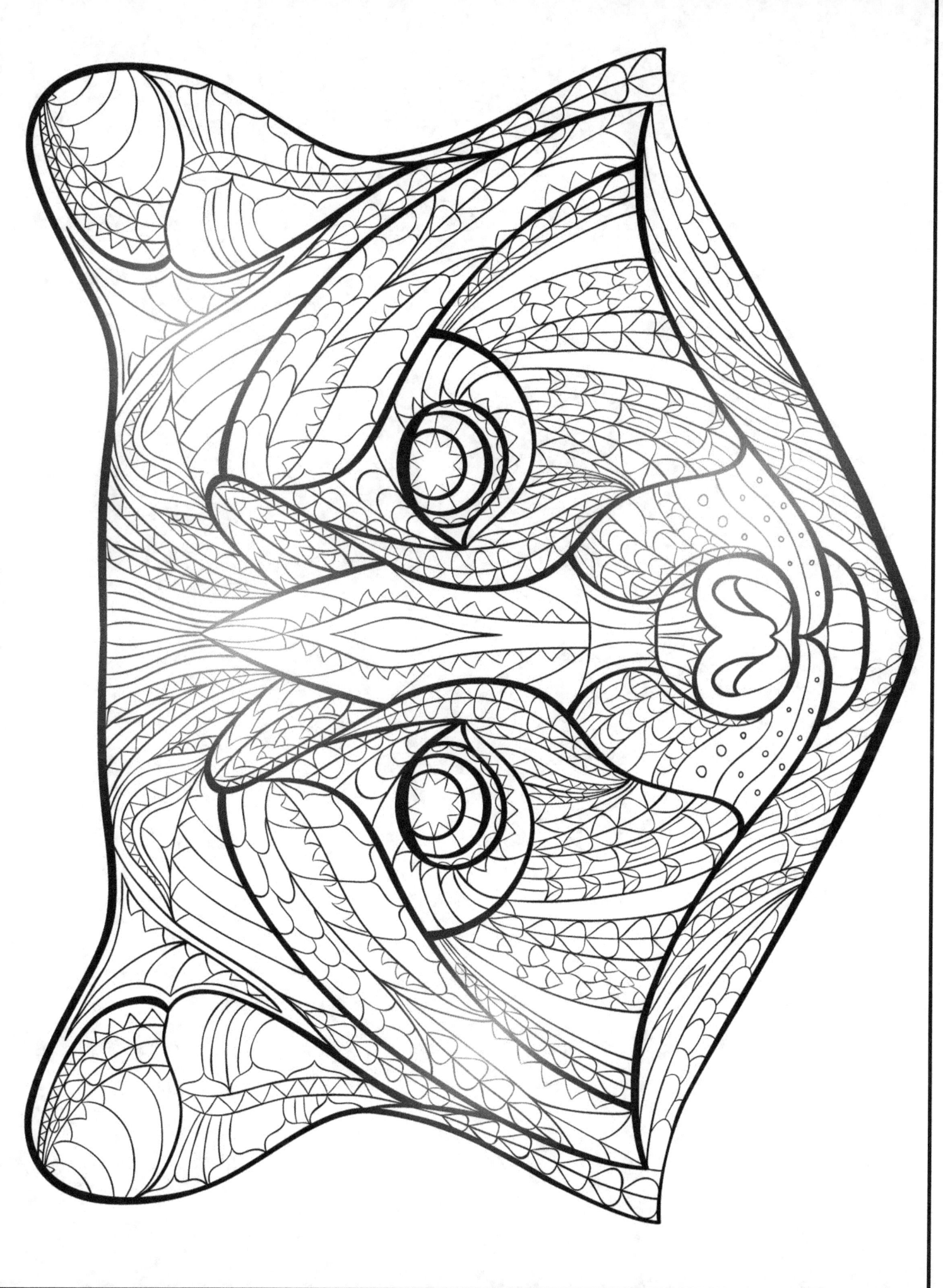

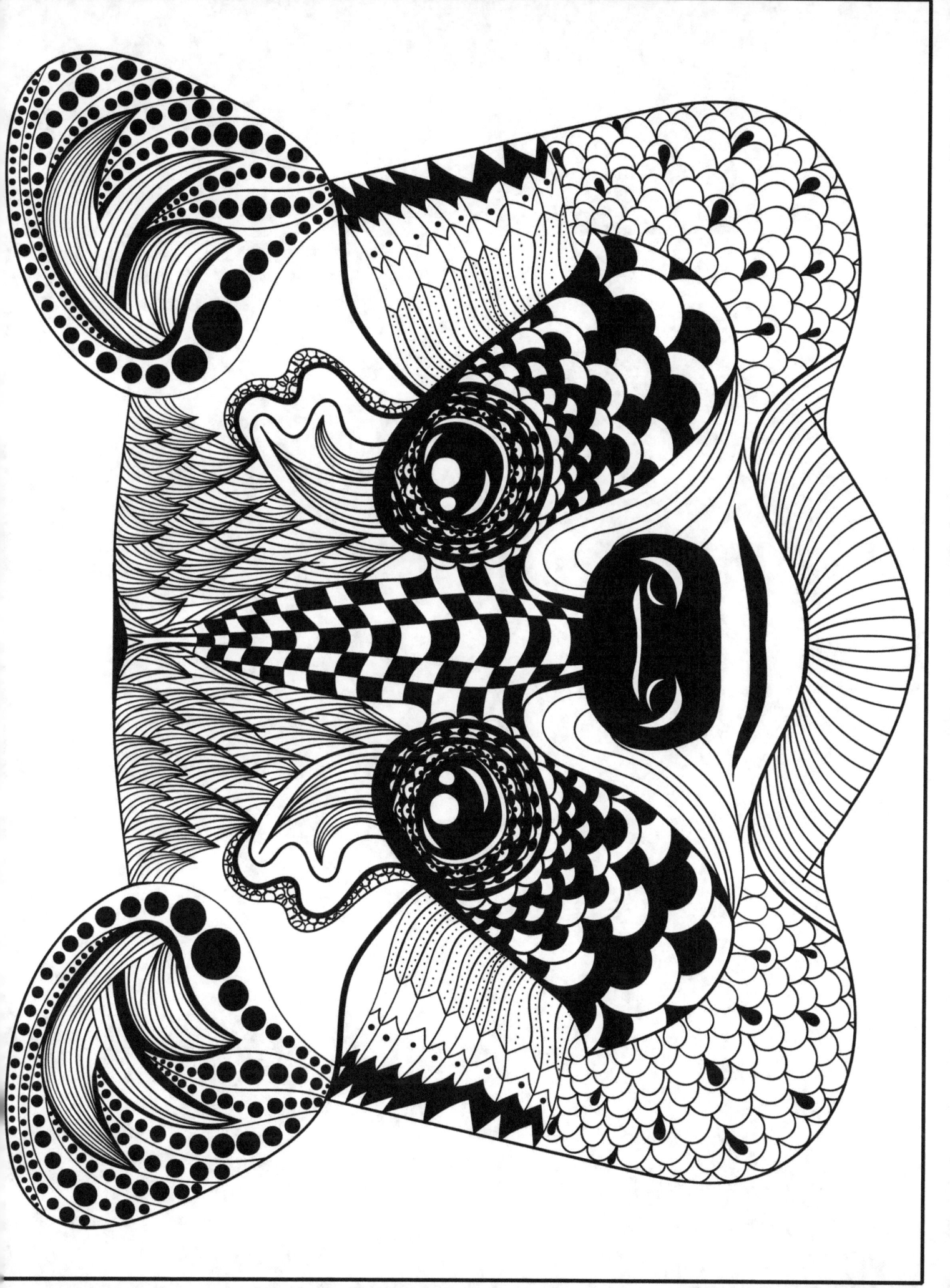

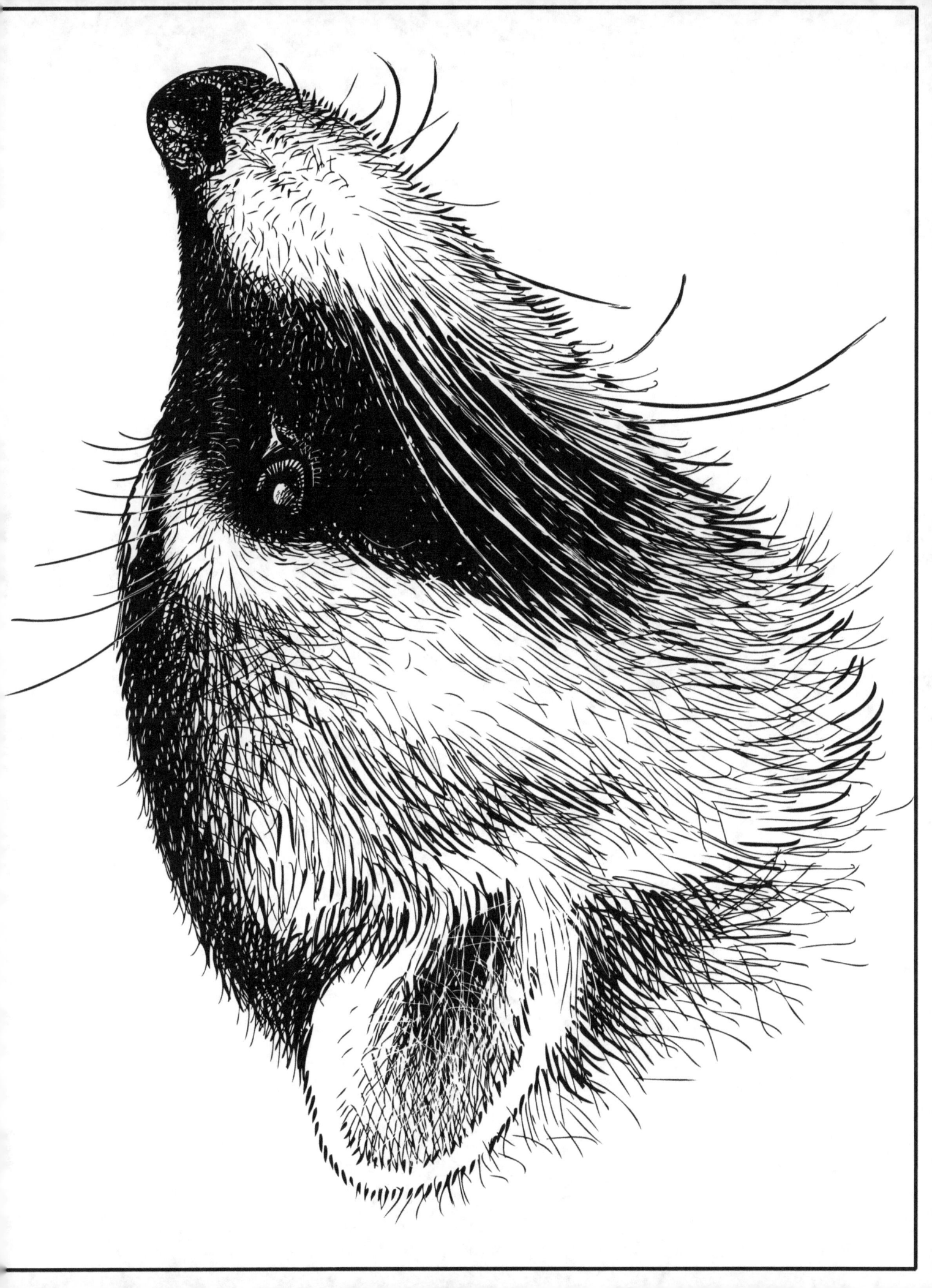

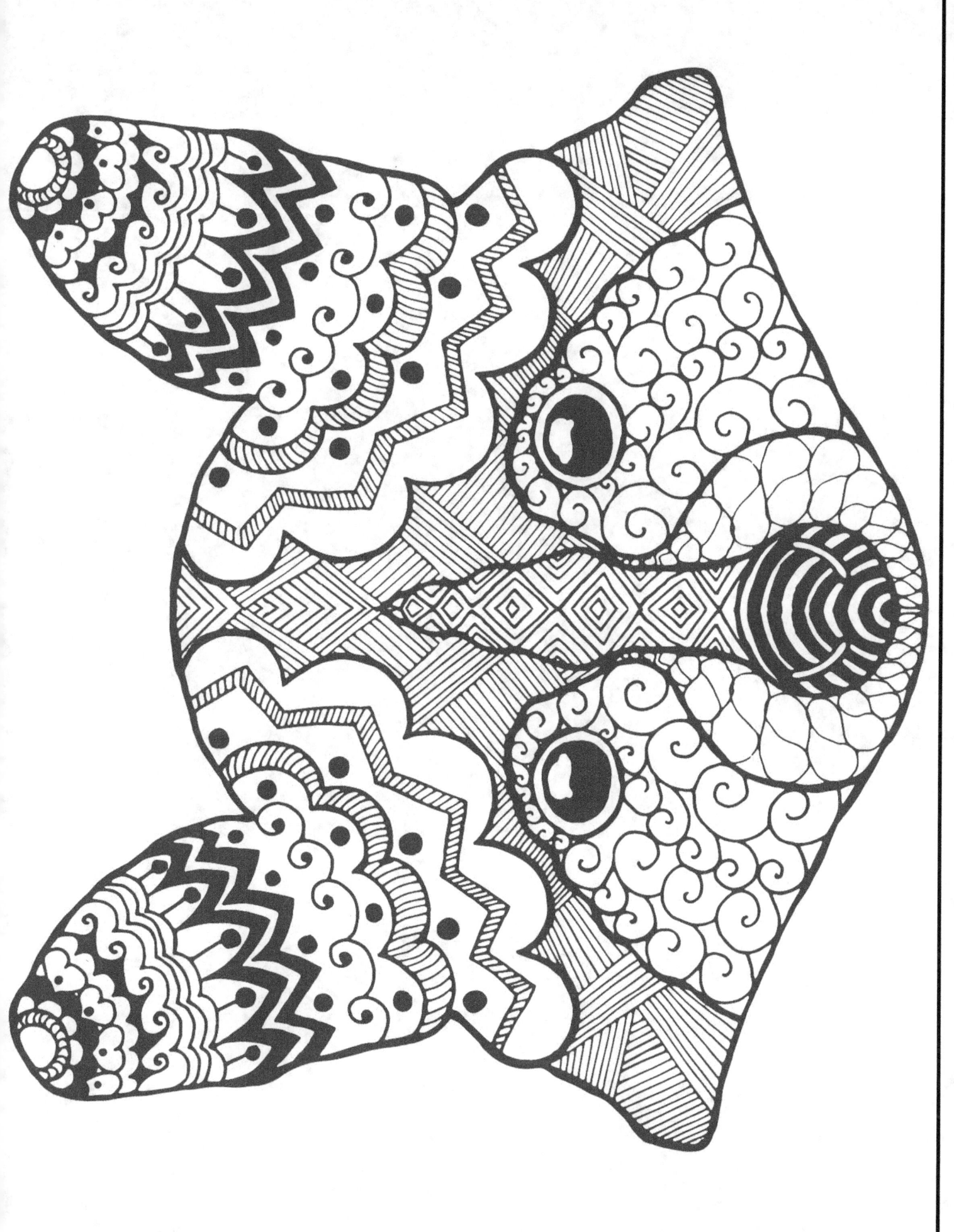

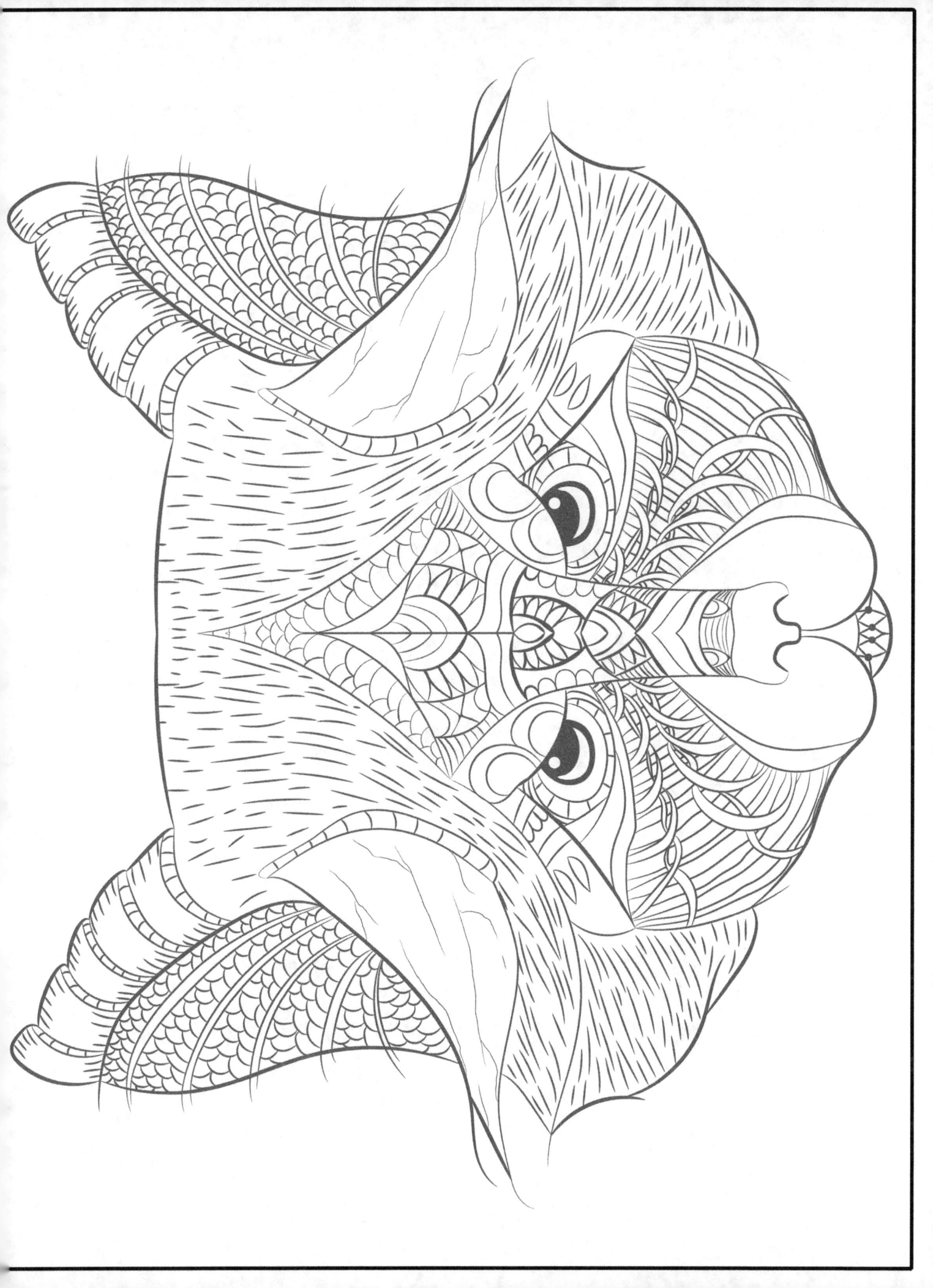

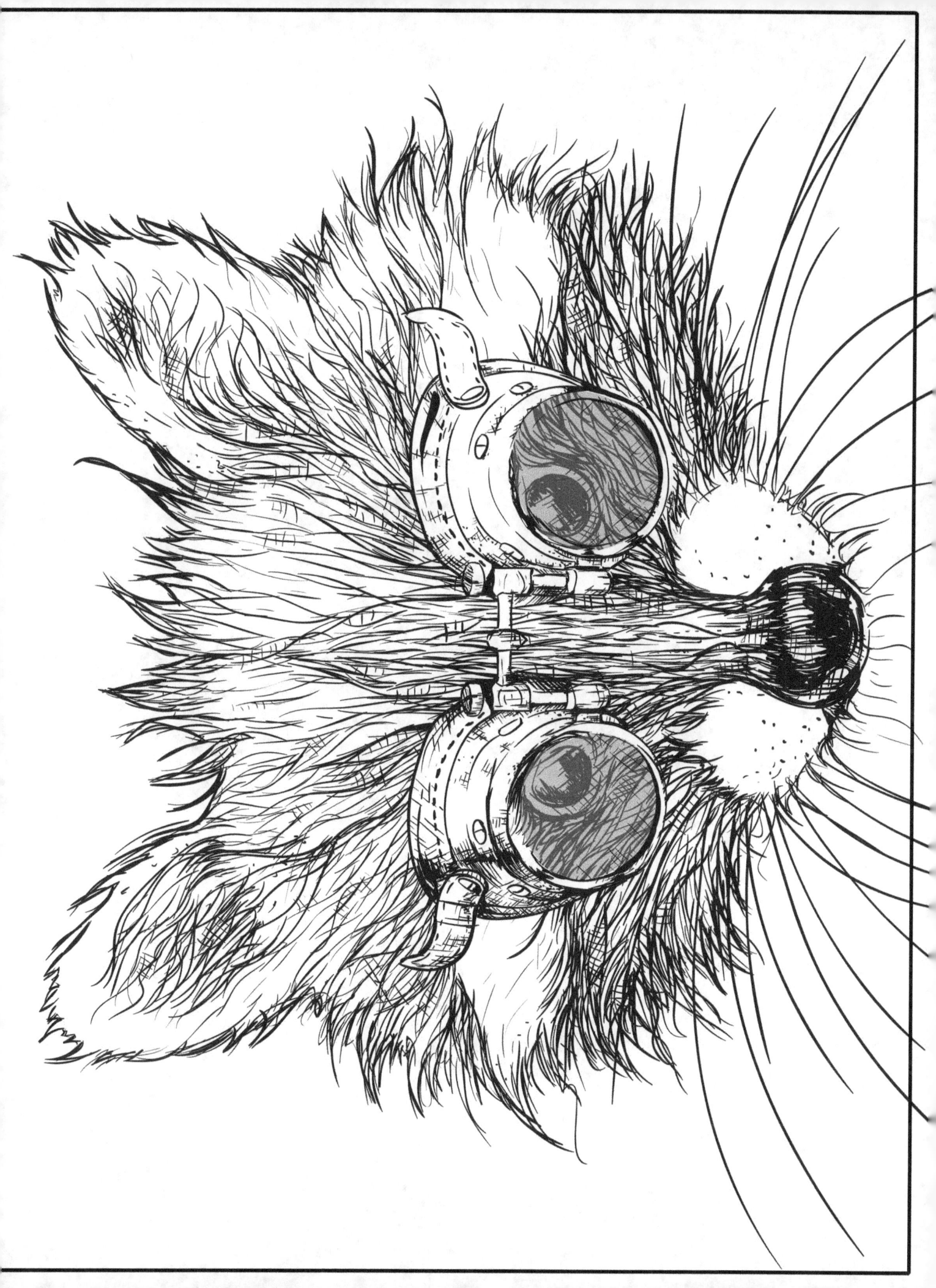

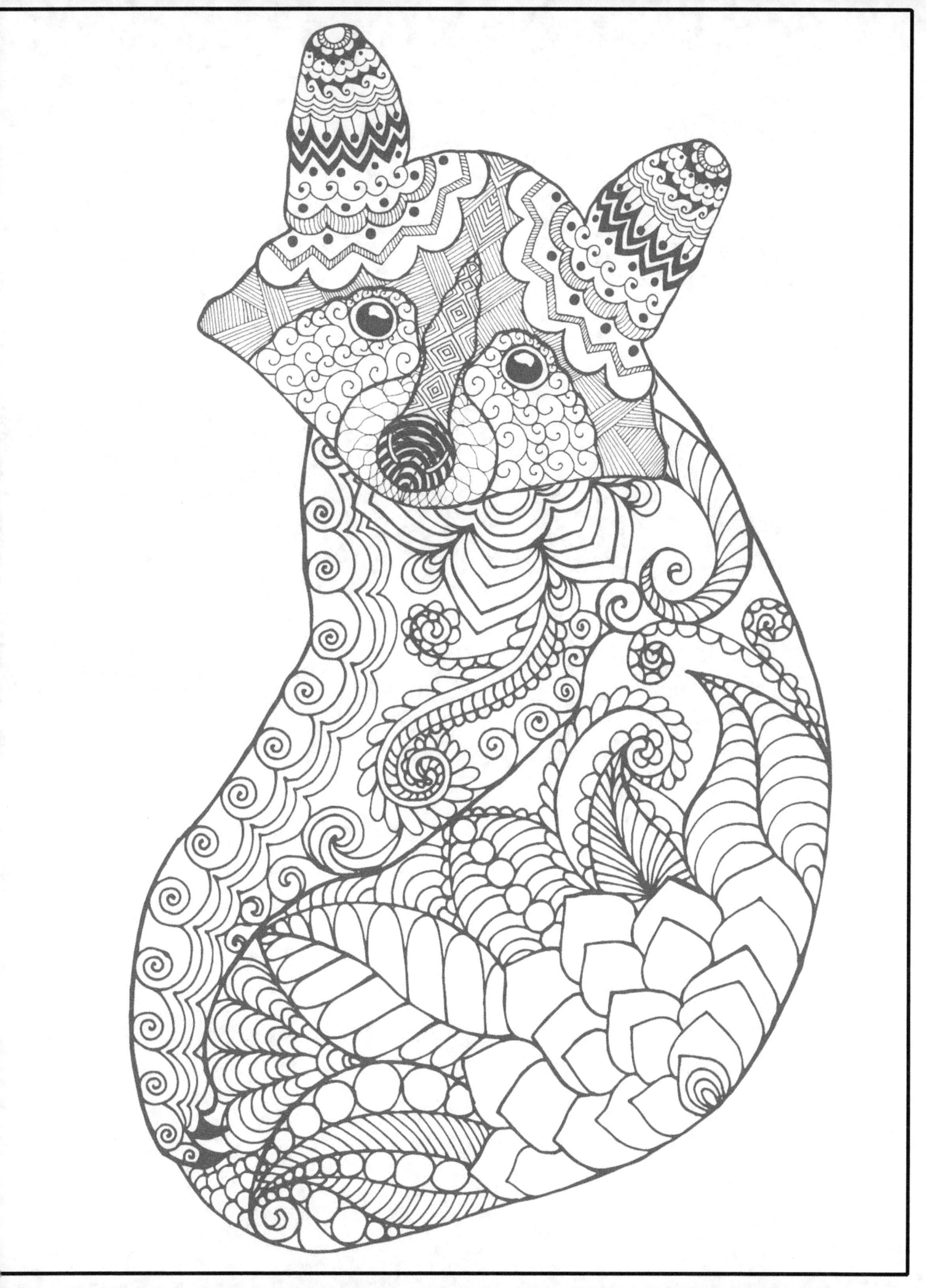

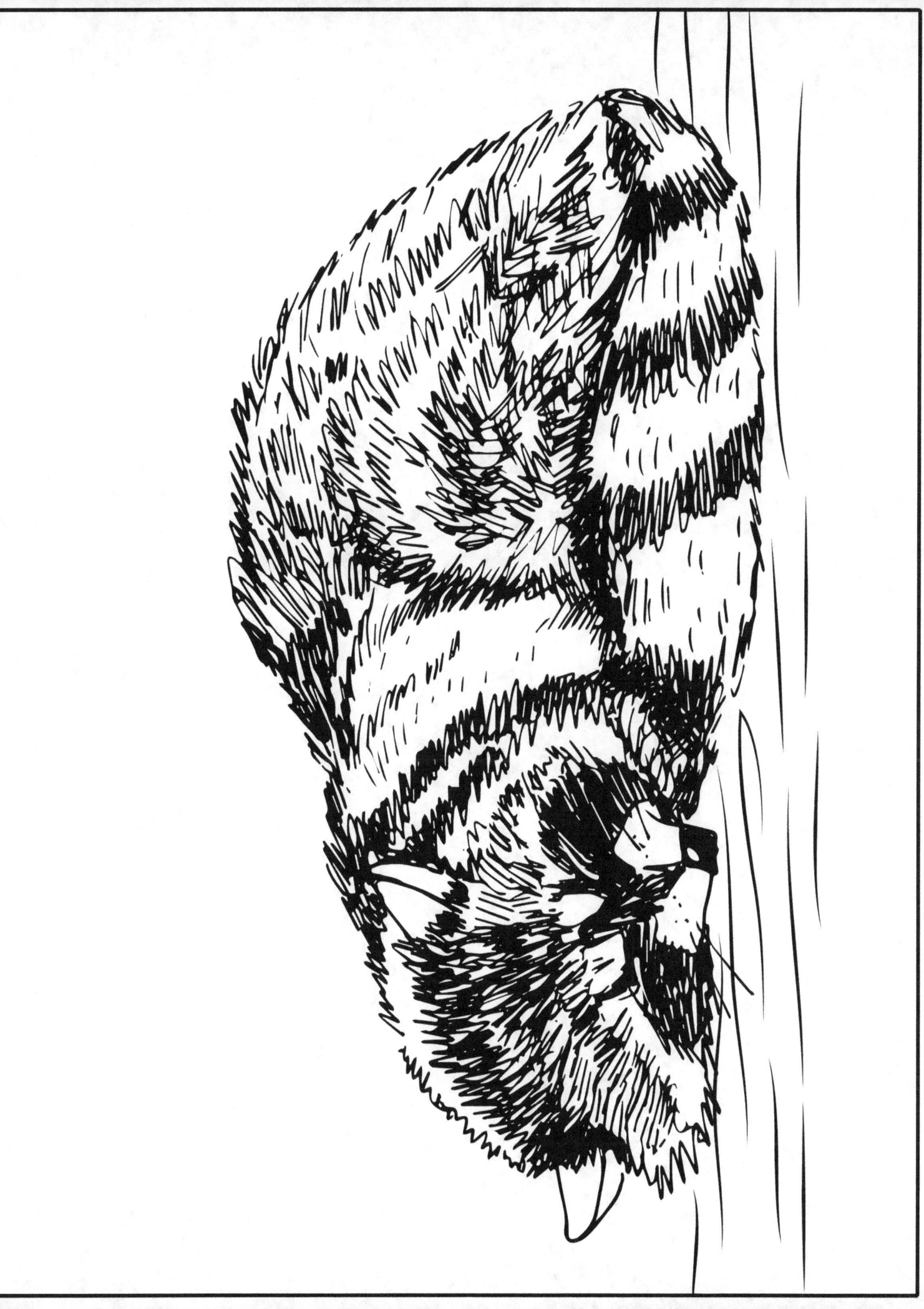

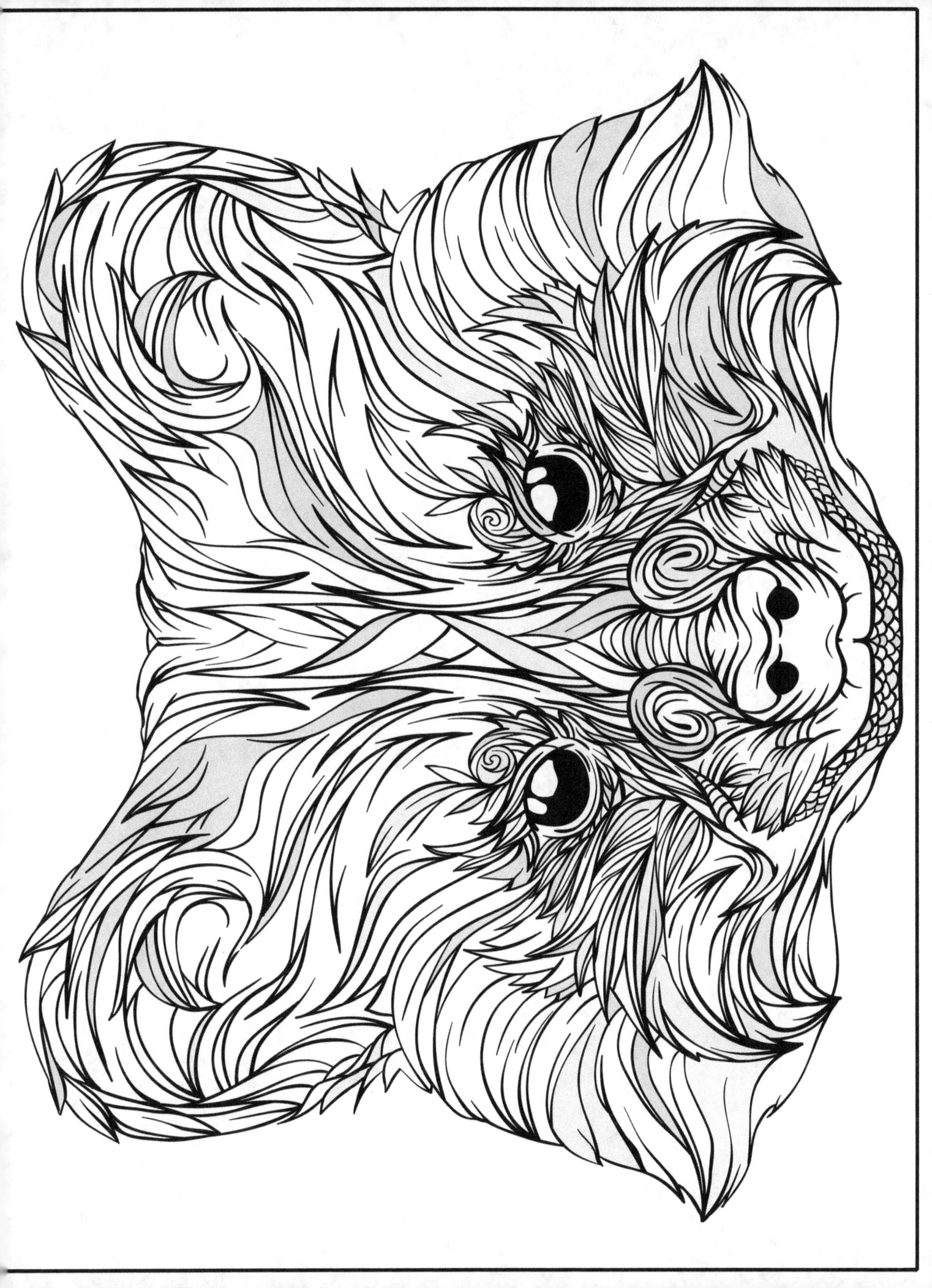

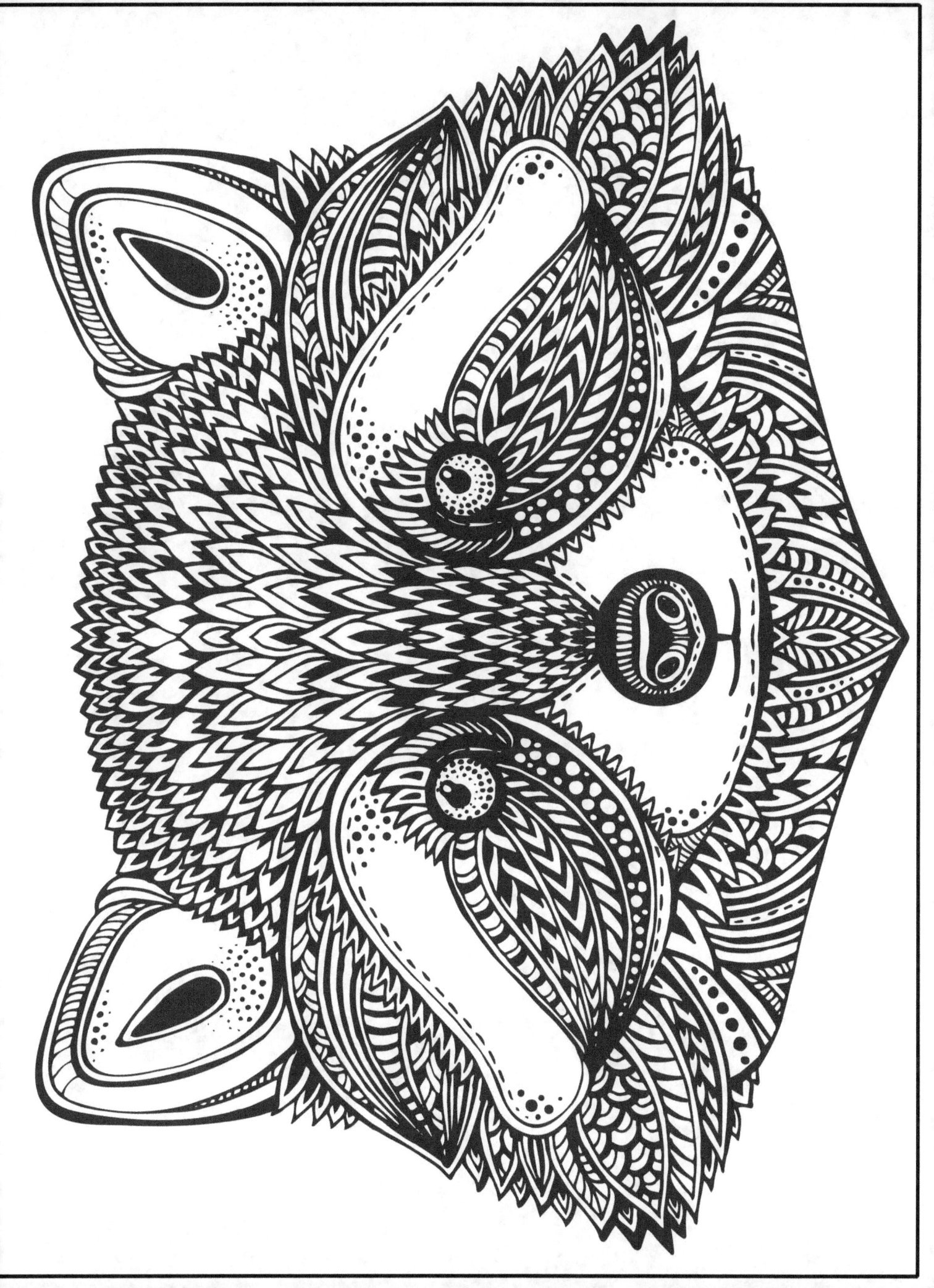

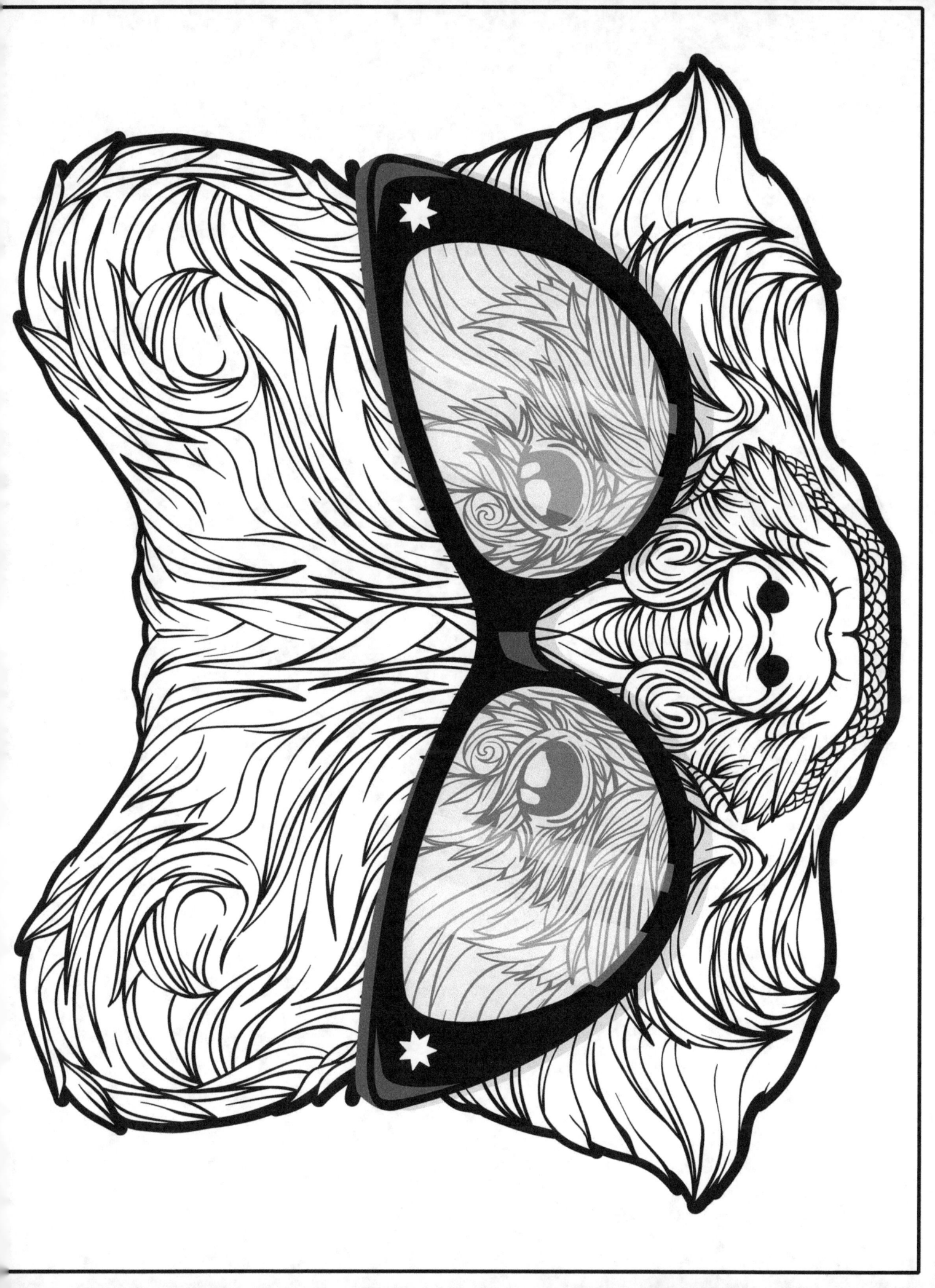

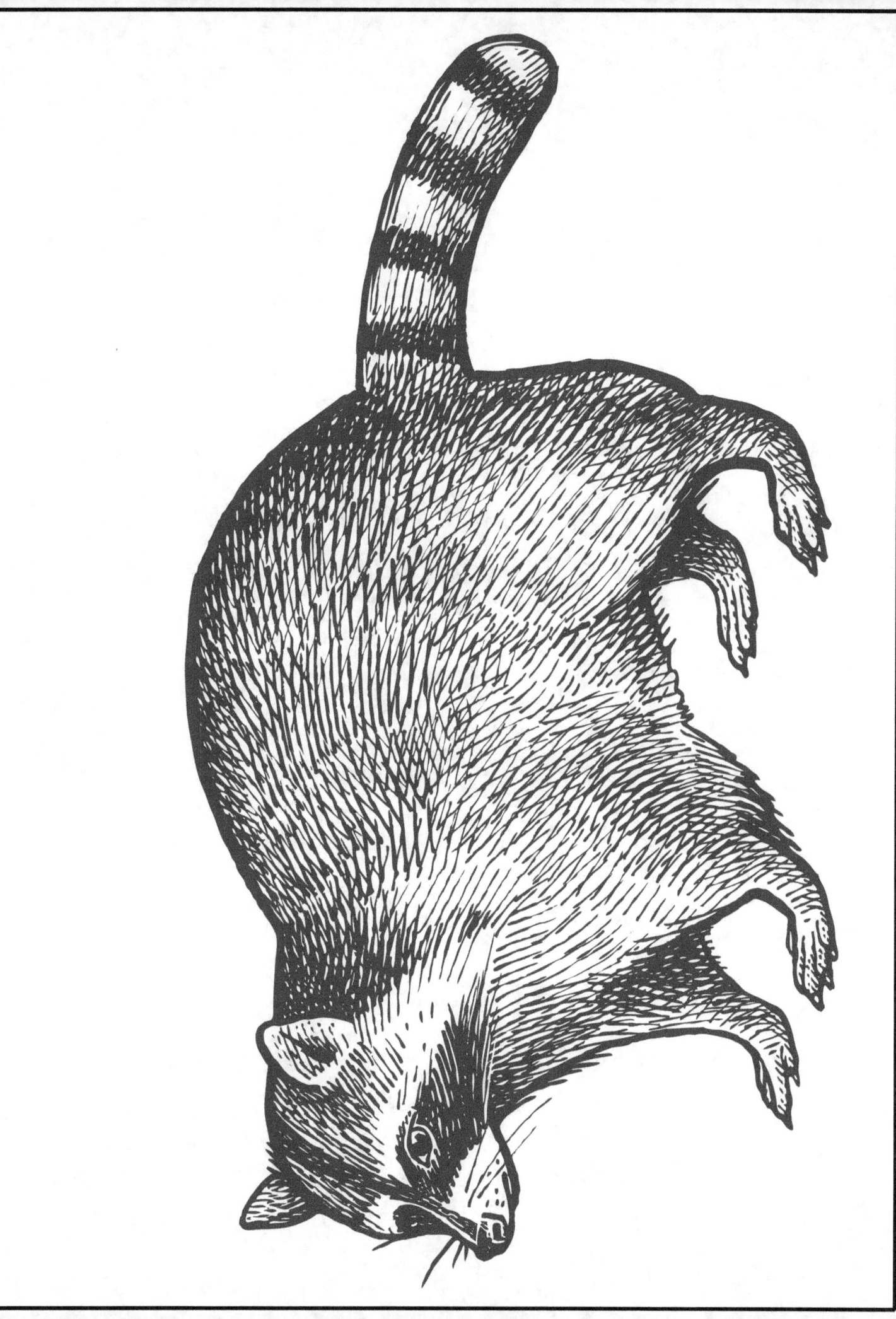

Join Us >> bit.ly/get_sample_free

- Get Free "Reviw Copies" of our New releases
- Exclusive offers and book giveaways
- More events from our community

Thank you

www.ingramcontent.com/pod-product-compliance
Lightning Source LLC
Chambersburg PA
CBHW081123180526

45170CB00008B/2978